HE

 OT

 O

 ONELY

STORY & ILLUSTRATIONS
BY:
ALEXANDER MELA

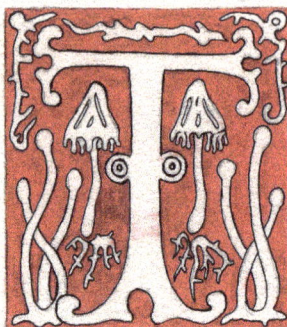 REE

Library of Congress Registration Number: TX 9-324-920
Hardcover ISBN: 979-8-218-29173-0
Paperback ISBN: 979-8-218-33666-0

Book design and illustrations by Dr. Alexander Mela

Name: Alexander Mela
Title: The Not-So-Lonely Tree

Description: 33 pages of watercolor illustrations | Audience: Ages 3-8 |
Summary: An ancient tree is lonely and isolated deep in a forest. But as
time goes on, it comes to find it is not so lonely after all, surrounded by a
whole hidden kingdom of fungal friends. This manuscript wraps complex
scientific concepts into a simple, watercolor-illustrated story. The tale is set
in an ancient kingdom with whimsical characters, meant to evoke a child's
curiosity and wonder. Backmatter includes a simple chart showing examples
of a few major characters from the story and estimated time scales of their
lineages, to put into perspective how closely related they are to each other
and animals.

Identifiers: ISBN: 979-8-218-33666-0 (paperback)
Subjects: Science & Medicine | Mycology | Botany | Microbiology | Art &
Illustration | Childhood Science Education

Deep in a forest,

There is a Kingdom that no one can see

No one knows of it

Not even The Lonely Tree

It's the oldest of its kind

It will be one thousand and one this spring

But there are ones older and wiser

An ancient queen and mushroom king

The tree looked around all day and all night

But to it's dismay

There were no others in sight

One morning, coming from the base of it's trunk

It smelled a strong, lovely, earthy funk

A shining Kingdom, just larger than a pea

Had been growing and thriving

Right beside the lonely tree

These are not plants

Not anything like me

But what wonderful friends

They could possibly be!

The tree looked closer

To see what's new

Then suddenly the fungi came into view

King Agaricus and Queen Button live in the forest too

They gather the mushroom roundtable every other June

The knights of the mushroom table

Come and gather round

To sing songs of the Kingdom Fungi

And to honor the crown

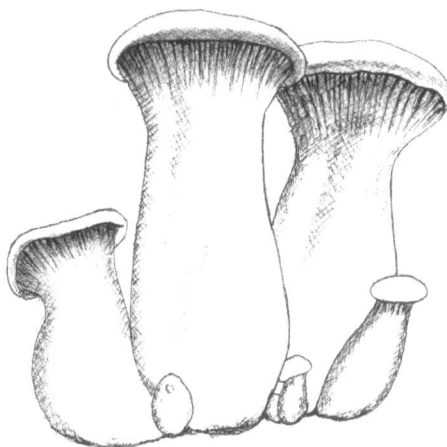

Queen Button
Agaricus bisporus

King Agaricus
Amanita muscaria

The Kingdom Fungi is HUGE

The party list is so long

That all their names must be sung out by song!

Violet Webcap
Cortinarius violaceus

The mushrooms are those that can live in plain sight

...Some are safe to taste...

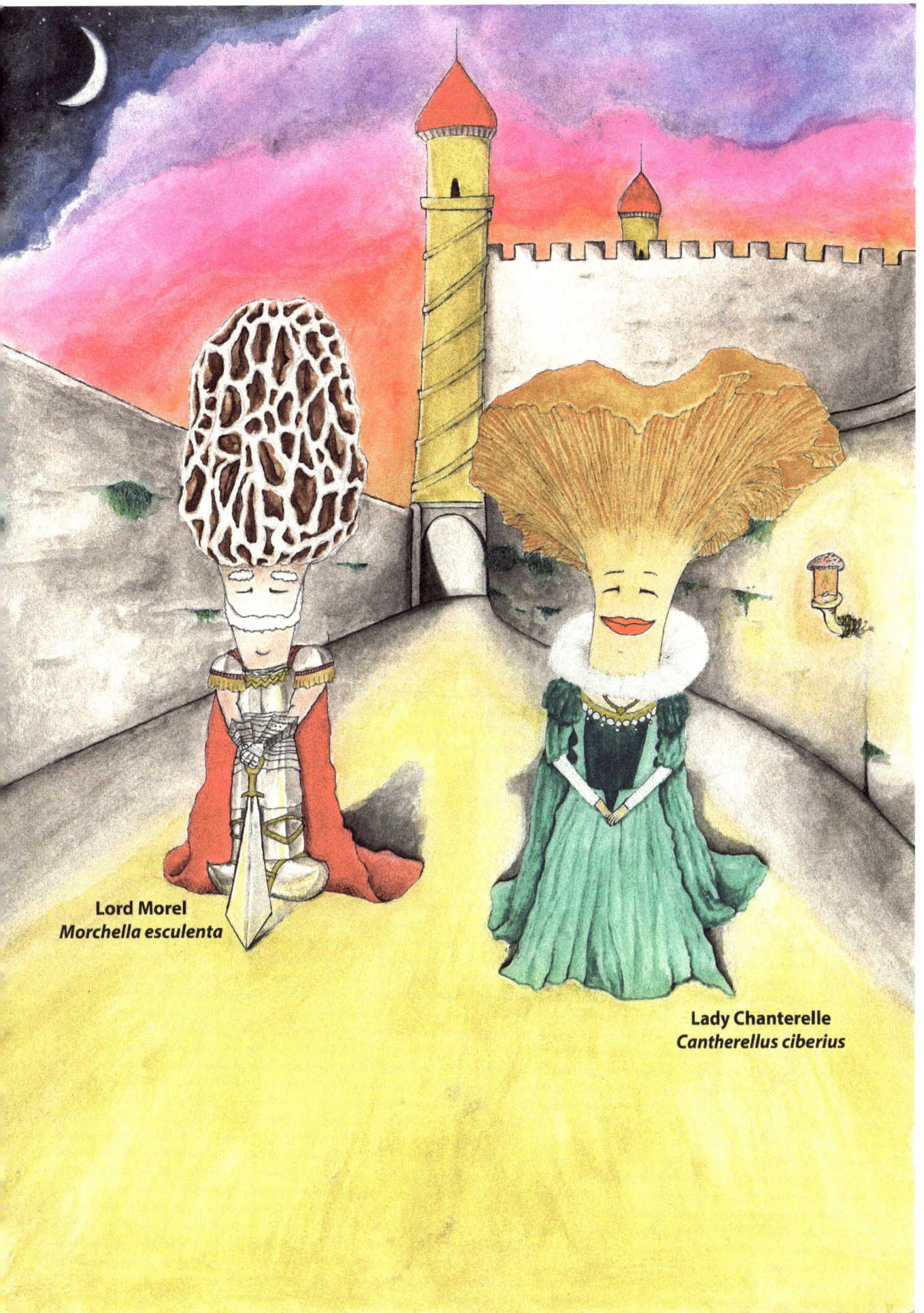

Lord Morel
Morchella esculenta

Lady Chanterelle
Cantherellus ciberius

...Some you simply should not bite

Lord Stinkhorn
Clathrus ruber

Lady Stinkhorn
Phallus impudicus

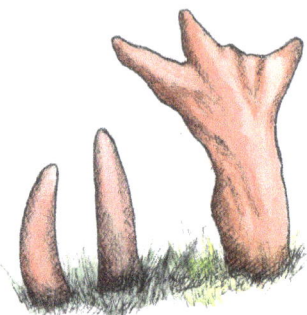

11

Bleeding Tooth Fungus
Hydnellum peckii

Destroying Angel
Amanita virosa

Inky Cap
Coprinus comatus

Panther Mushroom
Amanita pantherina

Devils Fingers
Clathrus archeri

From the tip of their cap

Down to their little stipe

Gilled or pored

Polka-dotted or striped

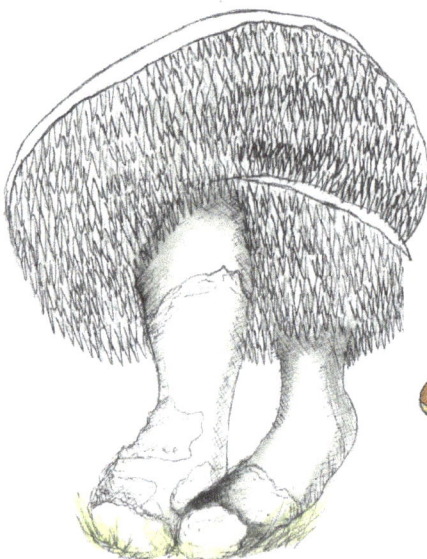

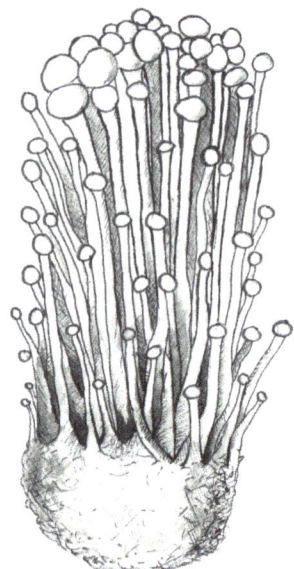

13

Artists Conk
Ganoderma applanatum

PORES

CAP

MARGIN

Chicken of the Woods
Laetiporus sulphureus

CAP / PILEUS
(pie·lay·us)

SPORES

GILLS / LAMELLAE
(lam·eh·lie)

RING / ANNULUS
(an·you·lus)

BULB

STALK / STIPE
(sty·p)

CUP / VOLVA
(vole·va)

Pixie's Parasol
Mycena interrupta

HYPHAE / MYCELIA
(high·fee) (my·see·lee·uh)

Puffball Mushroom
Lycoperdon periatum

14

"Many fungi could not be here today."

Said Madame Mould, in a most sincere way

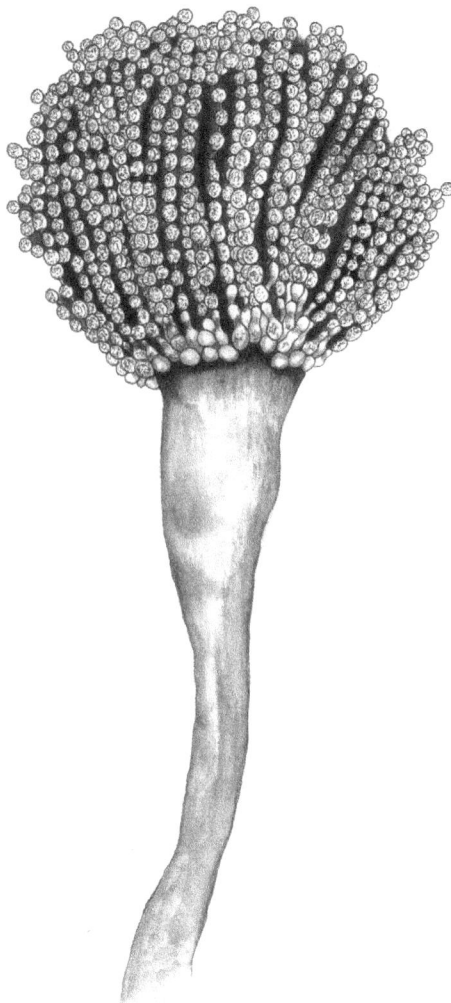

Madame Mould
Aspergillus nidulans

The Mycorrhizae (My-Kuh-Rye-Zee) are but a few

That could not part with ease

Constantly hugging the roots

Forever stuck to the trees

But do not worry

They always remain friends

The partnership goes both ways

It benefits both ends

The trees use the sun to make sugar in the leaves

And the Mycorrhizae feed their root system all that it needs

They connect the roots of not just one

They share food with short trees that never see the sun

NUTRIENTS

SUGAR

SUGAR

TREE FOOD

TREE FOOD

TREE FOOD

TREE FOOD

TREE FOOD

Arbuscular Mycorrhiza
Racocetra gregaria

The Mycorrhizae (My-Kuh-Rye-Zee) aren't the only fungi

That could not gather 'round

There is a clan called Truffles

That live deep underground

They come in all shapes and sizes

And are quite fragrant too

Pigs can't see past the soil

But can smell right through

Rare and prized by many

So they hide away deep

Under a grove of poplar trees

They take a nice...long...sleep

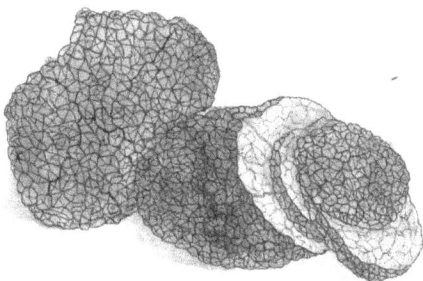

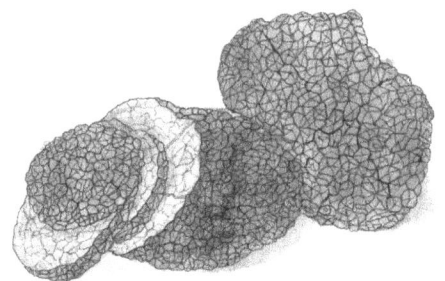

White Truffle
Tuber magnatum

From under the table a voice decrees...

Duke of Lichen
Xanthoria parietina

Dutchess of Lichen
Cladonia floerkeana

"There are fungi that can live anywhere they please!"

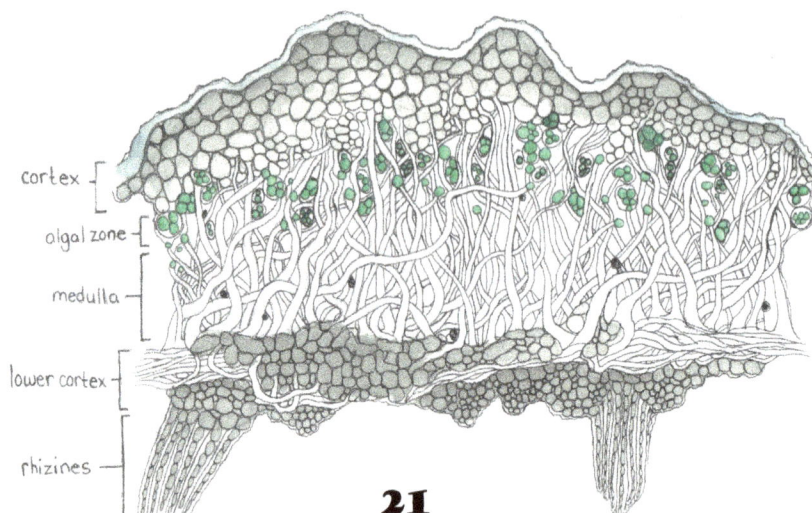

cortex
algal zone
medulla
lower cortex
rhizines

21

Old Man's Beard
Usnea longissima

"Lichen (Like-In) grow on cliff rocks
And the bark of tall trees
They make food from the sun
The rocks
Even air in the breeze"

Down in the fields

There's a really big hassle

The grass grows as tall

As the really big castle

The cows chew up the greens

But to digest the means,

The Chytrids (Kit-Rids) eat all the food

Just like hungry teens

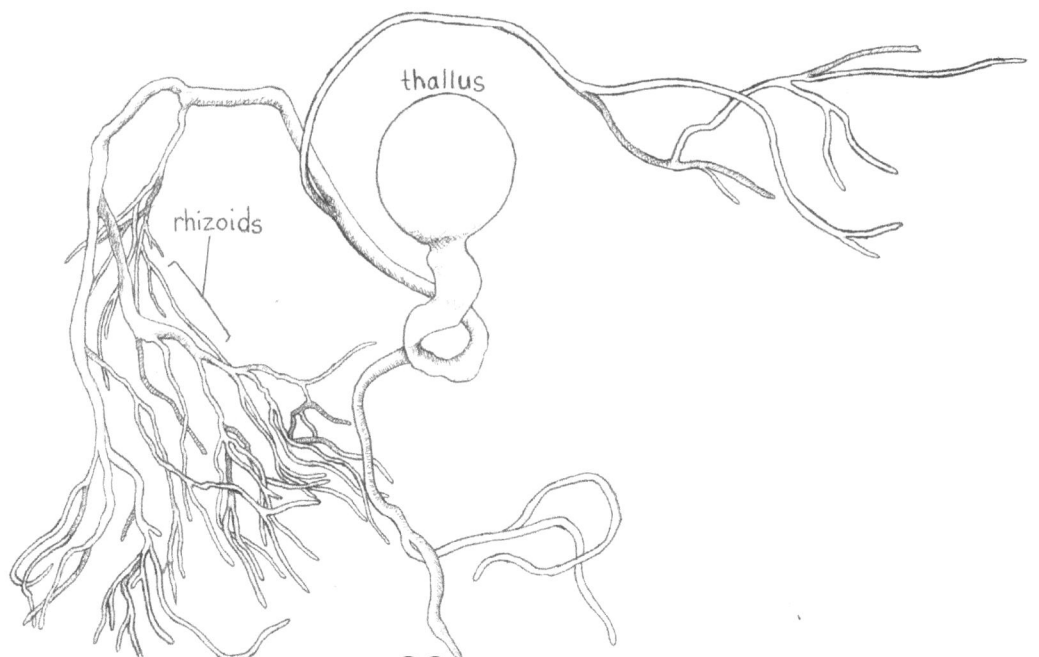

Rumen Chytrids
Neocallimastix frontalis

In the fungal Kingdom

Most live years untold

But those that grow old

Are never left in the cold

They're surrounded by good friends

The black bread mold

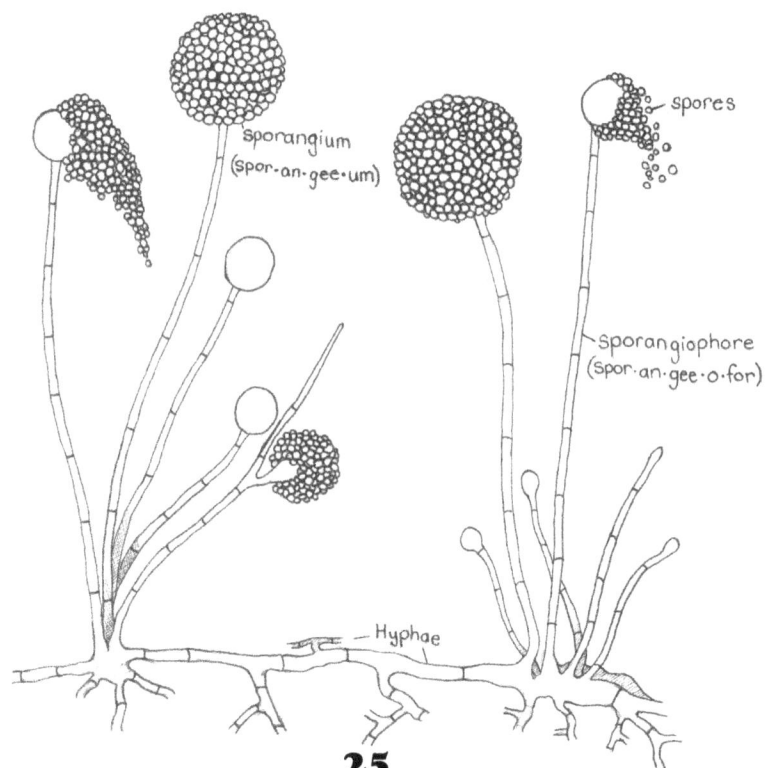

Black Bread Mold
Rhizopus stolonifer

26

"That's a wrap for today!"

The king shouted out in a most content way

"There are so many others

Who could not come to play

But do not be sad

We will meet after May

For a mushroom roundtable

Will gather real soon

To sing songs of the secret Kingdom

Just like this very June."

Wood ear
Auricularia auricula-judae

Jelly Fungus
Tremella fuciformis

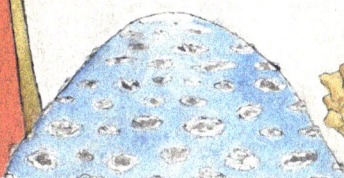

Jiggle
Wiggle
Jiggle
Wiggle

clack
click
clack

There's more than meets the eye

The fungi are here

They gather from everywhere

From far, from near

If you look close enough

You can clearly see

It isn't alone after all

That Not-So-Lonely Tree

Oldest Fungal Ancestor

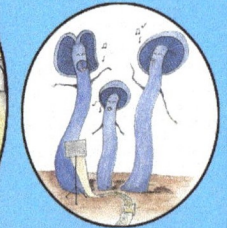

Chytridiomycota   Zygomycota   Glomeromycota   Ascomycota   Basidiomycota

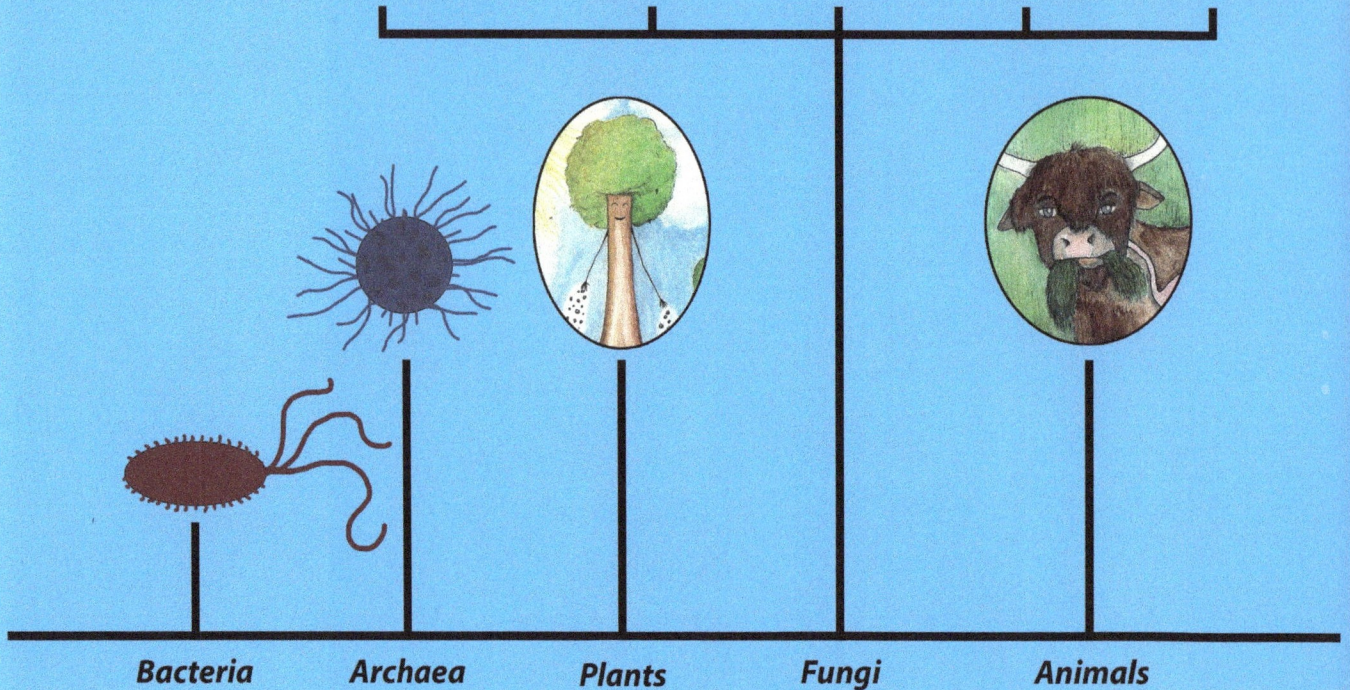

Bacteria   Archaea   Plants   Fungi   Animals

Animal and Fungi shared a common ancestor around 1.538 billion years ago

Plants are estimated to have diverged from Animals and Fungi 1.547 billion years ago

Bacteria and Archaea are estimated to have diverged ~3.4 billion years ago

The last common ancestor is estimated to have existed ~4.29 billion years ago

How Closely Related Are Fungi To Others?

www.ingramcontent.com/pod-product-compliance
Lightning Source LLC
Chambersburg PA
CBHW061305270326
41933CB00025B/3494